CH-47 Chinook

By Wayne Mutza

Color by Don Greer
Illustrated by Joe Sewell

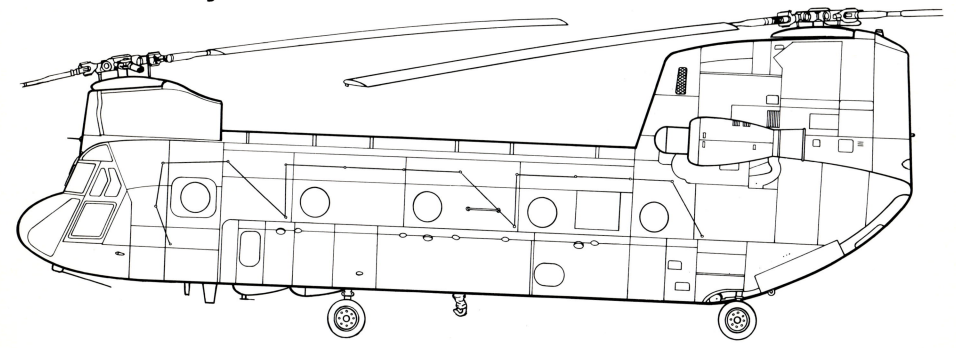

Aircraft Number 91
squadron/signal publications

An A/ACH-47A Chinook gunship of the 53rd AVN DET *Guns-A-Go-Go* of the 1st Cavalry Division based at An Khe, Vietnam, in the Summer of 1966.

COPYRIGHT © 1989 SQUADRON/SIGNAL PUBLICATIONS, INC.
1115 CROWLEY DRIVE   CARROLLTON, TEXAS 75011-5010
All rights reserved. No part of this publication may be reproduced, stored in a retrieval system or transmitted in any form by any means electrical, mechanical or otherwise, without written permission of the publisher.

ISBN 0-89747-212-8

If you have any photographs of the aircraft, armor, soldiers or ships of any nation, particularly wartime snapshots, why not share them with us and help make Squadron/Signal's books all the more interesting and complete in the future. Any photograph sent to us will be copied and the original returned. The donor will be fully credited for any photos used. Please send them to:

Squadron/Signal Publications, Inc.
1115 Crowley Drive.
Carrollton, TX 75011-5010.

Dedication

This book is dedicated to all CH-47 Chinook crews, past and present, who by their skill, dedication, and courage proved that the CH-47 was a formidable leader in the world of rotary wing aircraft. This book is also dedicated to Debbie, whose love instilled in me the spirit to complete this project.

Photo Credits

Art Babiarz
Madelyn Bush
Bob Chenoweth
George Ecker
Pete Harlem
Fred Koch
Bob Krenkel

Terry Love
Bob Mills
Bob Pickett
William Tedesco
Boeing Vertol
U.S. Army
Regina G. Burns
Library Tech/Archivist

A CH-47A Chinook of B Company, 228th ASHB, resupplies a remote mountain top base in Vietnam during 1968. The CH-47 was the primary medium lift transport helicopter used by U.S. Forces in Vietnam, moving both men and supplies under difficult and demanding conditions. (Boeing Vertol)

INTRODUCTION

In the years preceding the US involvement in Southeast Asia the U.S. Army dramatically expanded its Army Aviation Branch. The expanding tactical use of Army Aviation was reflected in the organization of Army combat field units which became centered around the squad size unit. This unit was considered the ideal size for helicopter air assaults. By early 1962, the Army was preparing to enter a new era, one centered around the concept of airmobile helicopter operations. With the implementation of the airmobile concept came the need for a large transport helicopter to move men and supplies within the combat zone.

In May of 1957, Vertol Aircraft (originally the Piasecki Helicopter Corporation) began designing, as a private venture, a twin turbine engined helicopter incorporating the same basic tandem rotor and transmission arrangement as the earlier Piasecki H-21. After twelve months, Vertol's prototype V-107 took to the air at Philadelphia International Airport on 22 April 1958. The aircraft featured a box shaped watertight fuselage for water landings, a rear loading cargo ramp, fuel tanks mounted in fuselage side sponsons, improved flight controls, and steel spar rotor blades mounted on vertical pylons at the front and rear of the fuselage.

The Department of the Army had announced plans to replace its radial engined CH-21, CH-34, and CH-37 troop carrier helicopters and had requested design submissions from a number of companies. Upon completion of a evaluation of the designs submitted by these companies, the design submitted by Vertol (which had merged with Boeing to become Boeing Vertol) was selected as the winner.

In July of 1958, the Army placed a contract for ten BV-107 prototypes under the designation YHC-1A. These were used to gain turbine helicopter experience and to evaluate the merits of the turbine helicopter as a medium transport helicopter under field conditions. The YHC-1A was considered by many Army officials to be too heavy for the assault role and too light for the transport role. After much internal debate over the proper size for the new transport helicopter, the Army finally decided to adopt a larger modified version of the BV-107, the Model BV-114, as its standard medium transport. The decision to adopt the larger BV-114 was directly related to service introduction of the Bell UH-1 Huey and the Army's plans for the expansion of air assault units. By pushing the Congress for funding for both the UH-1 and development of the Model 114, the Army hoped to accelerate the implementation of the airmobile concept.

The decision to proceed with development of the BV-114, under the designation YHC-1B, led to a reduction of the YHC-1A contract to three aircraft (these aircraft were later redesignated YCH-46Cs and were absorbed into the Navy). Five YHC-1B service test prototypes were ordered in May of 1959, while the three interim YHC-1As began flying in August. Design and construction of the first YHC-1B culminated with a roll-out of the first service test aircraft on 28 April 1961. This was joined later by a second prototype which made the first YHC-1B flight on 21 September 1959.

The service test aircraft emerged as tandem rotor helicopters with a strong family resemblance to the YHC-1A. They had a rather boxy shaped fuselage, with the fuel tanks, electrical systems, and quadricycle landing gear housed in two lateral pods on the fuselage sides. This arrangement was chosen to leave the cabin area unobstructed and allow maximum space for internal cargo. The cabin floor, lower fuselage compartments, and landing gear wheel wells are sealed to permit ramp-down water operations. The service test aircraft had a fuselage length of 51 feet, width of 12 ½ feet, and height of 18 ½ feet (to the rear rotor arc). The overall aircraft length, including the rotors, was 98 feet.

The tandem rotors were arranged to overlap, with the forward rotor canted 9° for-

One of three Boeing Vertol YHC-1As (Model BV-107) that were ordered by the Army in July of 1958 for evaluation of turbine tandem rotor helicopter performance. All three aircraft were painted overall Gloss White with International Orange trim. (Boeing Vertol)

ward of the vertical and the rear rotor disc canted 4°. The rear rotor pylon placed the rotor blade arc high above the hydraulically-operated rear cargo loading ramp, which allowed vehicles to be driven directly into the main cabin. The cabin is 30 foot long, 8 feet wide at floor level, and 6 ½ feet high along its entire length. The ramp doors are split with the upper portion retracting into the lower ramp section. The lower ramp also serves as an adjustable work platform and contains an escape hatch.

Power is derived from a pair of 1,940 shp Lycoming T55-L-5 turbine engines (uprated to 2,220 shp) which transmit power to the rotors through right angle gearboxes into a combining transmission gearbox mounted in the rear pylon. Vertical spoiler slots are contoured into each side of the forward rotor pylon. Along the upper fuselage walkway, between the pylons, are located hinged tunnel covers which provide access to the drive shaft and flight control linkage.

Boeing Vertol engineers incorporated a number of maintenance saving devices into the YHC-1B. Sight gauges for the oil-lubricated bearings and other vital components were designed to be accessible from within the main cabin, while integral flush steps, hinged cowlings, and stressed access panels double as work platforms. For removal/installation of major components, a removable crane can be attached to the base of each rotor pylon by means of a "plug-in" mount. These maintenance saving features were no accident. During the design phase, to constantly remind the Boeing-Vertol engineers of the requirement that the helicopter must be maintainable in the field, a standard Army mechanic's tool set was mounted to the wall of their work station.

The rotor heads are fully articulated assemblies with conventional flap and drag bearings. Movement of the control stick, left or right, tilts both rotors while fore and aft movement simultaneously decreases one rotor's pitch while increasing the other for diving or climbing. The foot pedals provide directional control by simultaneously tilting the forward rotors right and the rear rotors to the left. The foot pedals also operate the rear wheel brakes when taxiing with the forward wheels off the ground. Dual Stability Augmentation Systems (SAS) stabilize the aircraft in pitch, roll, and yaw. An integral T62 auxiliary power unit (APU) furnishes power for starting and systems checks, eliminating the need for ground support equipment. The APU's exhaust is vented through an opening at the base of the rear rotor pylon trailing edge.

The service test aircraft had an empty weight of 18,500 pounds and could carry up to 10,100 pounds internal or 16,000 pounds of external cargo. Flight tests revealed that the aircraft had a cruising speed of 90 knots with a maximum speed of 100 knots. The 564 gallon internal fuel capacity allowed a normal mission radius of 100 nautical miles.

A 16,000 pound capacity external cargo hook is mounted in a hatch in the main cabin floor. The hook itself is roller-mounted on a radius beam to prevent load motion from affecting flight attitude. The hatch door is stowed just to the rear of the opening against the aircraft's belly when the hook is in use. A 600 pound capacity rescue hoist can be mounted above the starboard cabin entrance and a floor-mounted winch, located at the front of the cabin, can pull up to 3,000 pounds through the rear fuselage opening. The cargo floor, which is covered with 5,000 and 10,000 pound capacity tie-down fittings, can support a load of 300 pounds per square foot. The forward dual-wheel landing gear is fixed and equipped with brakes, while the rear single-wheel landing gear units are movable and can swivel or be locked in place.

The service test aircraft carried a normal crew compliment of three, pilot (who sits in the right seat), copilot, and flight engineer (crew chief). For troop transport, side fuselage bucket seats are provided for thirty-three troops. If cargo requirements allow, another eleven seats can be fitted in the center cabin aisle. The cabin can also accommodate twenty-four stretchers for the medical evacuation role. For paratroop operations, a static line can be rigged along the cabin roof with the troops exiting from the rear ramp. The two-section cabin entrance door on the starboard forward fuselage comprises an upper jettisonable panel and a lower outward opening section that incorporates a built in boarding step. The upper section slides upward and into the cabin on rails. The cockpit doors, while not normally used, are jettisonable and each door contains a sliding window for cockpit ventilation.

The urgent requirement for HC-1s in Vietnam led the Army Aviation Test Board to compress the flight test program of the HC-1 and the service test aircraft were flown around the clock. During these tests one of the aircraft lost its rear rotor in flight and in the ensuing crash the aircraft was totally destroyed and all on board were killed. The accident led many officials, both in and out of the Army, to suggest that the Army was pushing the HC-1 program too hard and too fast and a rumor circulated that the entire program might be cancelled because the crash had been caused by a design fault. The Army investigation, however, revealed that the accident was caused by an improperly installed thirty-five cent bolt.

Before the YHC-1B evaluation was completed, the Army ordered a pre-production batch of five aircraft under the designation HC-1Bs. Soon afterwards, the Army assigned the aircraft the name Chinook, following the Army tradition of naming helicopters after North American Indian tribes.

During February of 1961 the Army ordered the Chinook into production with a contract for eighteen aircraft. This was followed by a further contract for twenty-four aircraft in December of 1961. During July of 1962, the HC-1B, along with the three YHC-1Bs, were redesignated under the Department of Defense revised designation system, with the YHC-1Bs becoming YCH-47As and the production aircraft receiving the designation CH-47A.

The U.S. Air Force acquired one of the YHC-1A from the Army after the Army decided to order the larger Boeing Vertol Model 114. Although carrying U.S. Air Force markings, this aircraft also carries a civil registration in Black on the fuselage underside. (Boeing Vertol)

This YHC-1B (59-4983) was the second of five service test aircraft and made the first flight for the YHC-1B during September of 1961. The aircraft was painted in Gloss White with International Orange panels and Black numbers and lettering. (Smithsonian Institution)

Development

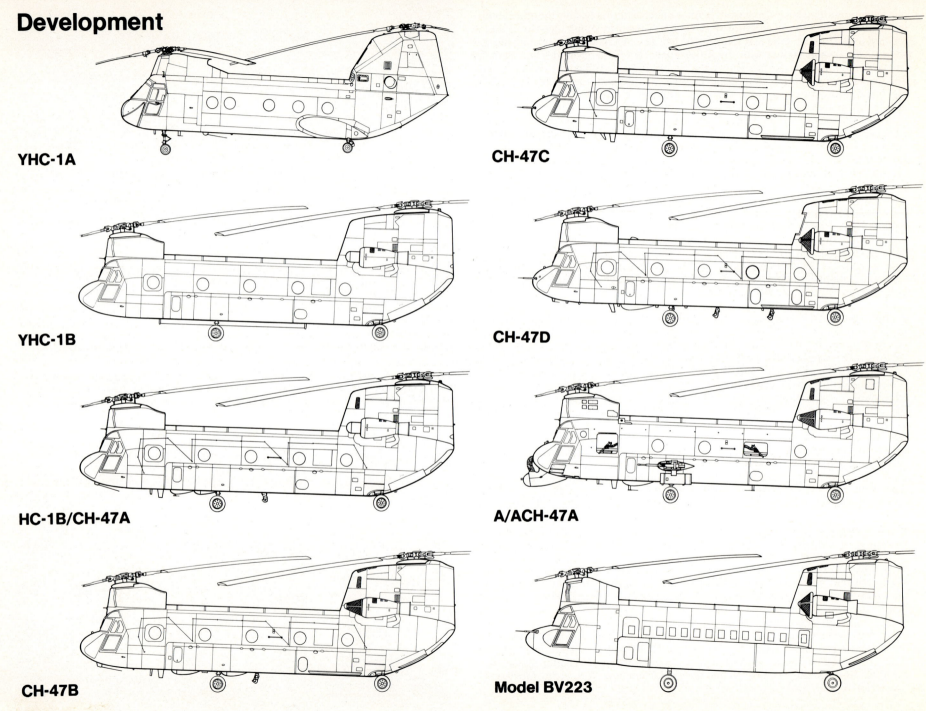

YHC-1A
YHC-1B
HC-1B/CH-47A
CH-47B
CH-47C
CH-47D
A/ACH-47A
Model BV223

CH-47A

The first production CH-47A Chinook was delivered to the Army on 16 August 1962. These aircraft were externally identical to the pre-production YHC-1Bs and differed only in internal equipment and avionics. One of the first units to receive the Chinook was the 11th Air Assault Division (which later evolved into the famed 1st Cavalry Division Airmobile). This unit used their Chinooks to refine the Airmobile concept and develop tactics for the Chinook in the air assault role.

Shortly after the Chinook was deployed to Vietnam during 1965, the engine air intakes were modified with protective dust/debris air filter screens to cope with the heavy dust which was common in Vietnam. These screens became standard equipment on late production CH-47As and were an optional feature that could be retrofitted to earlier aircraft.

Other changes brought about during the service introduction of the Chinook included, provision for an air-to-ground towing kit which allowed the CH-47 to tow large ground equipment and a rear landing gear 'kneeling' kit which reduced the Chinook's overall height by one and a half feet to permit storage on the hangar deck of Thetis Bay class aircraft carriers. Other field modification kits were produced so that Army maintenance units could modify their aircraft during overhauls. Two such kits were a winterization kit and a fuel cell self-sealing kit which protected the lower portion of the aircraft's fuel cells from bullet hits up to 7.62MM. The internal fuel capacity was also increased from 564 gallons to 621 gallons.

Early production CH-47As were powered by two 2,200 shp T55-L-5 engines; however, the need for additional power under the hot conditions in Vietnam led to the introduction of the 2,650 shp T-55-L-7B engine on late production aircraft. The engine change boosted the CH-47As maximum airspeed to 110 knots and allowed for an increase in maximum gross weight to 28,550 pounds (although operational experience in Vietnam showed that the CH-47A could be safely operated at weights up to 38,500 pounds). Chinooks operating in Vietnam's mountainous regions, under both hot and high conditions, however, were limited to a maximum payload of 7,000 pounds.

The CH-47A had a rate of climb of 1,590 feet per minute with a service ceiling of 9,200 feet. Hover ceiling out-of-ground-effect (OGE) was 7,300 feet. (However out-of-ground-effect is the ability of a helicopter to maintain a hover without the cushion of air produced by the rotor down wash striking the ground). Over the course of the production run a total of 354 CH-47As were built between 1963 and 1967 with the last production aircraft being delivered during early 1967.

The Chinook has established the lowest accident rate of any aircraft type in U.S. Army service. In Vietnam, the CH-47A Chinook quickly proved to be the Army's most important transportation vehicle — its ability to deliver large loads of troops and cargo into tight spots under fire and its outstanding mission availability earned it a reputation for dependability. The CH-47 quickly became a vital and indispensable tool in the Army's Airmobile concept.

The YHC-1B entered production under the designation CH-47A and was given the name Chinook. This early production Chinook is finished in Gloss Olive Drab with International Orange panels and high visiblity markings. All lettering, except for the White ARMY, is in Yellow and the cockpit frame is in Black. (U.S. Army)

The YHC-1B featured an easily loaded cabin compartment with a large cargo door/ramp at the rear of the cabin. Many of the inspection panels and soundproofing of this YHC-1B have been removed for tests. (Boeing Vertol)

This CH-47A carries the crest of the 11th Air Assault Division (forerunner of the 1st Cavalry Division) on the aft rotor pylon. The window outlined with a double Yellow frame is the cabin emergency escape hatch. (U.S. Army)

The CH-47 was developed to be easy to maintain. Large access panels in the rear rotor pylon allow for access to the combining transmission and rear oil cooler system. The raised fuselage spine is a hinged drive shaft tunnel cover which can be swung back to provide access to the drive shaft and flight controls. (Boeing Vertol)

This CH-47A parked on the ramp at Elmendorf AFB, Alaska during the Summer of 1967 was used for Arctic trials and was equipped with a ski landing gear. The skis went around the wheels and attached directly to the landing gear struts. (Norman Taylor)

The external cargo hook is mounted in an underfuslage bay on a radius beam, which prevents load motion from affecting the flight attitude of the CH-47. The external hatch slides to the rear along the fuselage underside when the hook is in use. (Boeing Vertol)

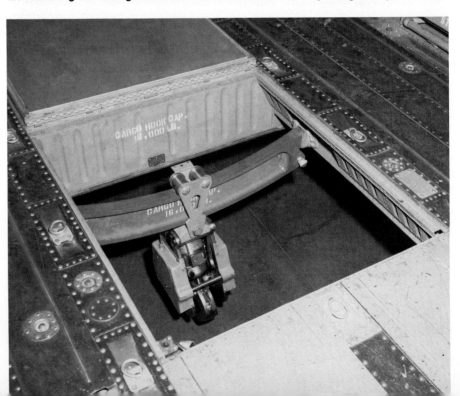

This CH-47A is undergoing external cargo lift tests at Fort Benning, Georgia, during 1962. The aircraft was assigned to the 11th Air Assault Division, which later became the 1st Cavalry Division. The Chinook is equipped with a rear view cargo mirror mounted under the cockpit. (U.S. Army)

Troops off-load from a CH-47A of B Company, 228th ASHB, 1st Cavalry at an airfield in Vietnam during 1967. The pierced steel planking (PSP) was a common sight at Vietnamese airfields and was commonly used for helicopter landing pads. (Tom Hansen)

This CH-47A on the ramp at McGuire AFB in June of 1979 is named *PUMPKIN*. The Red, White, and Blue tail emblem identifies this Chinook as being assigned to the 28th Aviation Battalion. The White lettering below the Orange pumpkin reads *DOC WATSON F.E.* (Steve Miller via Bob Pickett)

Engine Nacelle

CH-47A (Early)

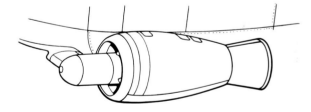

CH-47A (Late)

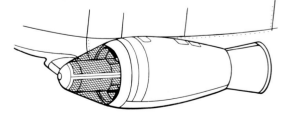

Specifications
Boeing Vertol CH-47A

Rotor Span	59 feet 1¼ inches
Length	51 feet
Height	18 feet 7.8 inches
Empty Weight	23,149 pounds
Maximum Weight	28,500 pounds
Powerplants	Two 2,650 shp Lycoming T55-L-5 turbine engines
Armament	Three M60D machine guns
Performance	
Maximum Speed	110 kts
Service ceiling	9,200 feet
Range	229 miles
Crew	Four

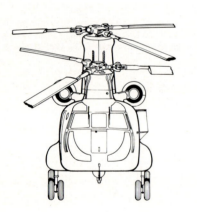
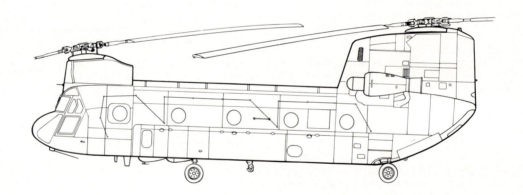
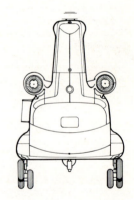
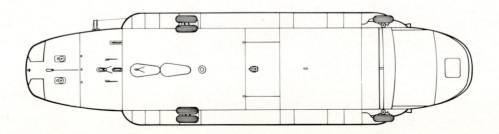

The Chinook has mounting points for a specially designed maintenance crane that enables maintenance crews to replace the Chinook's rotor head and transmission without the aid of outside ground support equipment. Once the rotor head and transmission are clear they will be swung out and lowered to the waiting maintenance stand. (Boeing Vertol)

The radio and electronic equipment compartment is located next to the passageway leading from the cargo/passenger compartment to the cockpit. The opposite side of the passageway has a similar comparment for the heating system and winch motor. (Boeing Vertol)

CH-47B

Experience gained with the CH-47A under combat conditions in Vietnam led the Army to consider changes to the aircraft to give it greater lift and performance under tropical conditions. One of the five pre-production YHC-1Bs (s/n 59-4984) was re-engined with two 2,850 shp T55-L-7C turbine engines fitted with modified rotor blades, becoming the prototype for an improved CH-47 under the designation YCH-47B.

The modifications to the YCH-47B included a redesigned rotor system with new blades that featured cambered leading edges. The forward blades were slightly longer than on the CH-47A, with the front rotor arc being increased from 59 feet 1¼ inches to 60 feet. The new blades were also stronger and featured a strengthened steel main spar and honeycomb trailing edges, in place of the of the aluminum main spar and fiberglass trailing edges used on the CH-47A.

The rear rotor pylon was changed with the pylon fin cap and trailing edge being squared off and the bulged APU exhaust port now becoming a flush mounted port. The rear fuselage was modified with two airflow strakes being added to the lower rear fuselage and the ramp door to improve lateral stability.

The YCH-47B successfully completed testing and was ordered into production under the designation CH-47B. Deliveries of production CH-47Bs began on 10 May 1967. Because of the demands of the Vietnam war, the tempo of production was increased and by the end of 1968 a total of 108 machines had been delivered to the Army for duty in Vietnam.

The additional power of the T55-L-7C engines on the CH-47B significantly enhanced the Chinook's performance under combat conditions. The CH-47B was faster, being capable of a cruising speed of 140 knots and a maximum speed of 155 knots. The rate of climb was 2,080 feet per minute and the CH-47B could hover OGE at 10,500 feet. With a full fuel load the CH-47B had a mission radius of 95 miles.

The additional power also increased the Chinooks cargo carrying capability. The CH-47B could carry a maximum internal payload of 15,900 pounds or an external load of up to 20,000 pounds on a strengthened external cargo hook. The aircraft's maximum allowable gross weight was 33,000 pounds, some 4,500 pounds more than the CH-47A.

This early production CH-47B still carries the factory construction number on the rear fuselage sponson in White. The rear view cargo mirror mounted on the framing under the cockpit was later deleted when the crew chief became responsible for monitoring external hook loads. (Boeing/Vertol)

A new production CH-47B on a pre-delivery test flight. The contractor number B383 on the sponson in White will be removed before the aircraft is delivered to its final destination. The aircraft is painted in overall Flat Olive Drab. (Boeing Vertol)

Rear Rotor Pylon

CH-47A   **CH-47B**

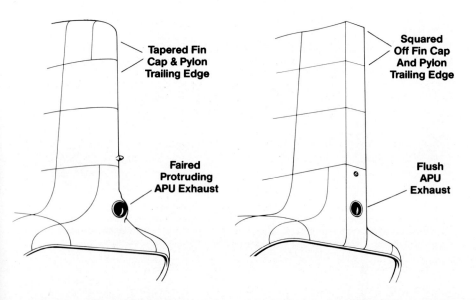

A CH-47B taxies across a lake to demonstrate the Chinook's capabilities to operate from water. The CH-47 can safely operate in open water with waves of up to thirty feet. (Boeing Vertol)

The rear ramp also serves as an adjustable work platform for servicing the rear transmission and auxiliary power unit (APU). Two identification features of the CH-47B are the strakes added to the lower fuselage and ramp and the squared off rotor pylon trailing edge. The outlet on the lower portion of the rotor pylon is the APU exhaust. (Boeing Vertol)

With the hinged cowling and access panels open, the entire engine of this CH-47B of the 271st Aviation Company is exposed for inspection and maintenance. The access panel for the transmission and rotor head is stressed to also serve as a work stand. (Boeing Vertol)

The absence of the pitot tubes on this CH-47C is because Army censors have deleted the aircraft serial number and unit markings. The aircraft was slated for duty in Vietnam during 1968 and the Army was sensitive about revealing the identity of units deloying to Vietnam. (Boeing Vertol)

A CH-47B of the 178th ASHC, Boxcars, prepares to lift an external sling load of ammunition from a truck at Chu Lai, Vietnam for movement to a forward area. All but one of the cabin windows have been removed to provide gun firing ports for troops. (William Kaminski)

A CH-47B of the 132nd Aviation Company delivers a fuel bladder to waiting M113 ACAV armored personnel carriers at Landing Zone (LZ) STINSON in October of 1969. The crewman in the open escape hatch is manning an M60D machine gun. (U.S. Army)

Rear Fuselage Strakes

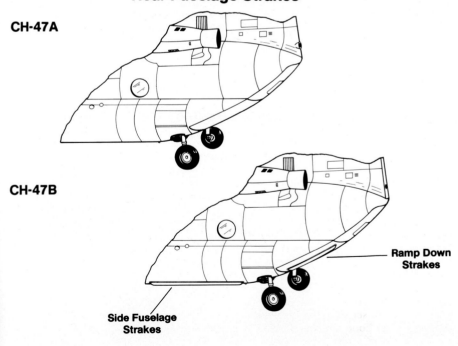

CH-47C

During the mid-1960s the Army issued a requirement for a helicopter that could lift a 15,000 pound external payload over a distance of 34 miles. This performance was beyond the capability of the CH-47A and CH-47B, which could lift the weight, but not carry it the required distance. The Boeing Vertol Company, however, thought that they could improve the CH-47 enough to meet the requirement.

To meet the Army requirement, Boeing Vertol upgraded the CH-47B with more powerful engines, an uprated drive shaft system, and dual automatic flight control systems. Other systems improvements included a dual Stability Augmentation Systems (SAS), vibration dampers, and a crash-worthy fuel system that could withstand hits from ground fire up to .50 caliber machine gun rounds. Internal fuel capacity was increased from 1,042 gallons to 1,130 gallons, permitting a range of 149 nautical miles.

The first CH-47C, was externally identical to the earlier CH-47B, except for the pitot tube, which was relocated from the its position on the front of the forward rotor pylon to a position on the front of the nose below the cockpit. Separate tubes were now fitted for both the pilot's and copilot's instruments. The CH-47C also introduced larger engine air intake dust/debris screens mounted on the front of each engine nacelle. The CH-47C made its first flight on 14 October 1967, with the first production examples being delivered to the Army on 30 March 1968.

The CH-47C was powered by two 3,750 Lycoming T55-L-11 turbine engines in place of 2,850 shp T55-L-7C engines used on the CH-47B. The increase in power allowed internal cargo loads to be increased to 24,000 pounds. External sling load capacity was also increased with loads of up to 22,700 pounds now being common over short distances. Thanks to the more powerful engines, some units operating at high altitudes in Vietnam (such as the Central Highlands) were now able to haul cargo loads of up to 10,000 pounds, where previously they had been restricted to loads of no more than 7,000 pounds.

Performance for the CH-47C was also improved thanks to the more powerful engines. The additional power gave the CH-47C a cruising speed of 150 knots, some ten knots faster than a CH-47B while maximum speed increased from 155 knots to 165 knots. The rate of climb for the CH-47C increased from 2,080 feet per minute to 2,880 feet per minute and the hover out of ground effect was raised from 10,500 feet to 14,700 feet.

Like its predecessor, early production CH-47C Chinooks were equipped with cambered rotor blades, however, late production machines were fitted with new rotor blades developed by the National Aviation and Space Administration (NASA). These blades were high-lift cambered blades made of fiberglass, aluminum, and steel, which were lighter than the steel and aluminum blades used on the CH-47B. Most early production CH-47Cs were retrofitted with the new blades at Army maintenance facilities as they went through scheduled overhauls.

The CH-47C also had the provision for carrying four bladder type internal fuel tanks in the cargo compartment. These ferry tanks gave the CH-47C a ferry range of over 1,000 nautical miles (at 10,000 feet). With these tanks, Chinooks became capable of cross Atlantic flight, enabling the CH-47C to be self-deployed to Europe instead of having to be airlifted inside USAF cargo aircraft. To demonstrate this self-deployment capability, four CH-47Cs made a record breaking flight across the Atlantic as part of Operation *NORTHERN LEAP* held during early 1979. The Chinooks took off from Fort Carson, Colorado and flew non-stop to Heidelberg Army Air Field, Germany, setting a new helicopter distance record.

The CH-47C served as the test bed for a series of flight tests intended to evaluate a new external cargo hook configuration known as the Tandem Cargo Suspension System (TCSS). The TCSS involved fitting two additional external cargo hooks to the CH-47C, which allowed suspension of heavier loads under the CH-47. The stress of the load was now spread to two hook points, instead of a single point, permitting greater weights to be carried and greater speeds to be flown, since the load is less likely to swing. This system was later standardized as part of the CH-47D program.

A total of 270 CH-47Cs were built before production was shifted to a more advanced version of the basic Chinook.

A new production CH-47C awaits delivery to an Army unit on the Boeing Vertol ramp. The long ventral strakes fitted to the lower rear fuselage and ramp helped improve lateral stability, especially in a hover. (Boeing Vertol)

Big Windy a CH-47C rests in its revetment at Phu Cat, Vietnam on 4 November 1970. The name was carried in White high on the rear of the aft rotor pylon. (Norm Taylor)

This CH-47C was fitted with a water spray rig for a series of tests to determine the effects of icing on various helicopters. These tests were conducted in Alaska during the late 1970s. The spray rig was painted Orange to make it visible to the helicopters flying behind the CH-47. (Aviation Archives via Terry Love)

The starboard forward fuselage of the ice test CH-47C was adorned with a variety of unit and company decals. Additionally a caricature of an ice laden Chinook was painted on the left side of the crew door. (Terry Love)

The water spray rig was deployed below the Chinook and put out a thin spray of water which quickly froze on aircraft surfaces in the cold Alaskan sky. The UH-1 Huey behind and below the Chinook is a test subject. (U.S. Army)

Nose Development

CH-47A/B   CH-47C

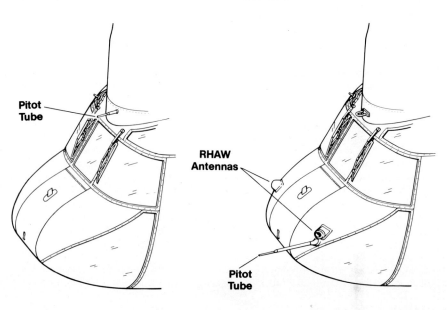

A CH-47 is off-loaded from a C-133A transport. To transport the CH-47 the rotors, rotor heads, and engines were removed, however, these were usually loaded on the same aircraft and the entire CH-47 could be transported to a distant location aboard a single large transport. (Boeing Vertol)

Dust/Debris Filters

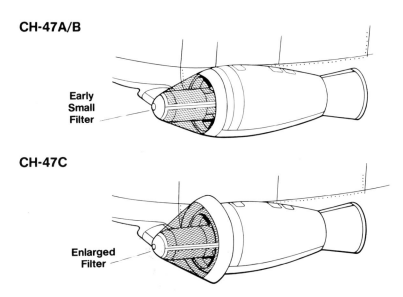

CH-47A/B — Early Small Filter

CH-47C — Enlarged Filter

A badly weathered CH-47C of the 213th Aviation Company is unloaded at Geissen Army Air Field, Germany, during exercise *Reforger* during 1981. The Chinook carries a Black cat unit marking on the front of the forward rotor pylon and on the sides of the rear rotor pylon. (U.S. Army)

The instrument panel of a CH-47C had a complete set of flight instruments for both the pilot and co-pilot, while the center console has the engine instruments. The directional control pedals under the instrument panel also controlled the wheel brakes. The pilot's thrust control handle is to the right of the lower center console. (Boeing Vertol)

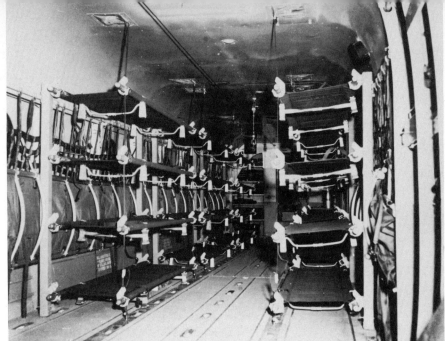

The Chinook fuselage could easily accomodate twenty-four litters for the medical evacuation role. The stretchers were easily installed since the normal troop jump seats simply folded up against the cabin sides. (Boeing Vertol)

GERONIMO, a CH-47C of the 20th Aviation Company carried the name in White on the rear rotor pylon during 1976. A number of post-Vietnam war Chinook companies have retained their Vietnam era unit markings. (Jerry Geer via Terry Love)

A CH-47C recovers a QF-102A Delta Dagger drone target. Chinooks play a major part in the recovery of target drones from remote areas of the gunnery ranges of the southwest United States. (via Pete Harlem)

Vietnam

The CH-47 was first committed to action in Vietnam during 1965 and quickly proved to be dependable, highly adaptable to the adverse weather conditions in Vietnam and capable of hauling tremendous loads into hostile areas, often under fire, and surviving. The 1st Cavalry Division deployed with their organic CH-47 Battalion to Vietnam in September of 1965, and by year's end, the division's Chinooks had amassed thousands of flight hours and had successfully recovered over 100 downed aircraft.

During that year, the Department of Defense ordered production of the CH-47 doubled to meet the expanded requirements of Army airmobile units in Vietnam. Appropriately nicknamed the 'Hook' by the troops in the field, the CH-47 quickly proved to be an invaluable tool for artillery movement and heavy supply lifts. The Chinook was seldom used for its intended role, that of an assault troop carrier; that mission being undertaken primarily by the smaller and more maneuverable UH-1 Huey. When the Chinook did transport troops into combat, however, lifts of up to seventy-five Vietnamese soldiers on a single flight were not uncommon. One of the primary missions of the CH-47 in Vietnam was the emplacement of artillery batteries at remote mountain fire bases and keeping these bases well supplied with ammunition.

The normal crew complement of a CH-47 in combat consisted of a pilot, copilot, flight engineer/crew chief (who also doubled as a gunner), and gunner. To prepare the aircraft for combat the crews removed the rear cabin windows (and sometimes other windows as well) to create gun ports for use by the onboard troops. The standard armament fitted to CH-47s in Vietnam was a pintle-mounted 7.62MM M60D machine gun in the port side escape hatch opening and a second M60D on a swing out mounting in the forward starboard crew door. Designated the XM-24, this armament subsystem contained mechanical stops to prevent the gunners from inadvertently firing into the rotors or fuselage during the heat of combat. The M60s were equipped with bipods and could be easily removed from their quick-release mountings for use as ground weapons in the event the CH-47 was shot down.

To prevent damage to the rotors and engines, a canvas bag was attached to the starboard side of each gun to catch ejected shell casings and links. A 200 round ammunition box was fitted to the port side of the M60, although these were usually removed in favor of much larger containers, depending on the unit policy or crew preference. Some units rigged flexible ammunition chutes (borrowed from 7.62MM Minigun mounts) which were fed by large ammunition boxes lashed to the cargo floor. The XM-24 subsystem had a rate of fire of 550 to 600 rounds per minute (rpm) and an effective range of 1,500 meters.

Although not as commonly used, the XM-41 subsystem incorporated an M60D or .50 cal. machine gun mounted on a pintle mount on the rear cargo ramp, firing out of the open ramp. This system was first tested during mid-1967 and saw some limited use as an area fire suppression weapon when exiting a hot landing zone.

In addition to the weapons, the radio equipment fitted to Vietnam based CH-47s was also enhanced. FM radios, used to communicate with ground units, were installed and the aircraft were modified with wire antenna mounting spikes on the port fuselage side. A number of aircraft also carried a pair of FM whip antenna mounts on the nose. These communications modifications were made to all variants that saw service in Vietnam (CH-47As, CH-47Bs and CH-47Cs) and were performed at the unit or depot level.

At the height of the war, some twenty-two Chinook units were in operation, performing a variety of missions in support of the war. One of the vital missions was the recovery of downed aircraft, known as *Pipesmoke* missions. Chinooks were used to salvage so

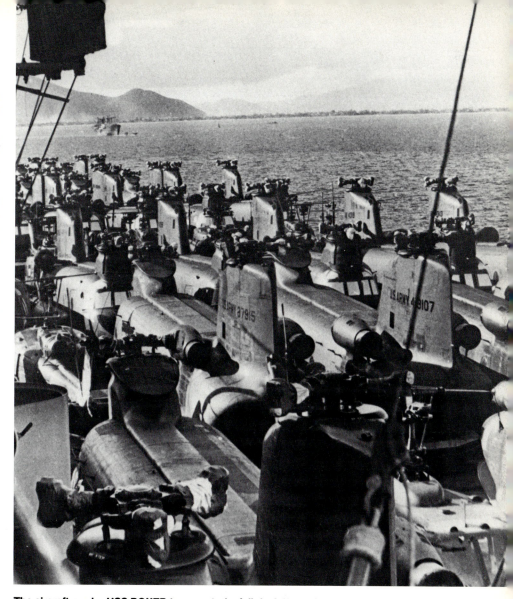

The aircraft carrier USS BOXER transported a full deck load of 1st Cavalry Division CH-47 Chinooks to Vietnam during 1965. During the transit the rotor heads were sealed to protect them from the effects of salt air and spray. (U.S. Army)

many downed aircraft that the CH-47 became the Army's primary recovery vehicle. Over the course of the war, CH-47s were credited with the recovery of 11,500 disabled aircraft worth more than three billion dollars. Equally important was the Chinook's use in the civic action effort which often required the rapid movement of an entire village to a safe location. During one such airlift, a Chinook lifted 147 Vietnamese and their possessions in a single flight!

'Hooks' also proved invaluable for supplying fuel to remote areas. For these operations the 3,500 pound, 500 gallon rubberized fuel cells, known as blivets, were carried as external cargo, suspended from the CH-47's external cargo hook.

Other vital missions performed by Chinooks in Vietnam included medical evacuation (MEDEVAC), engineer support, paratroop drops, air-to-ground towing, and troop insertion and extraction. One effective method of inserting and retrieving troops in jungle areas made use of rope ladders suspended from the rear cargo ramp. This method was rather unsettling for the troops, who dubbed the method the "Jacob's Ladder;" it also tested the pilot's skill because it required maintaining a hover for long periods of time, often in hostile areas and under fire.

During the Tet Offensive, with troops engaged in the battle to retake Hue in February of 1968, Chinooks were used to set up an aerial supply bridge between the battlefield and cargo ships just offshore, in what was the first ship-to-shore combat resupply effort in Army history.

Most Vietnam-based Chinook units were under the operational control of the 1st Aviation Brigade, being organized into Assault Support Helicopter Companies (ASHC) each with a complement of approximately sixteen aircraft. Other Chinooks operated under their parent commands, the 101st Airborne Division and the 1st Cavalry Division (Airmobile). At the height of its involvement in 1969, the 1st Aviation Brigade possessed 311 CH-47s.

Typical of GI inventiveness was the adaptation of the CH-47 as a 'bomber'. Throughout Binh Dinh Province the Viet Cong had built up a large number of fortifications and an extensive tunnel system that could withstand anything but a direct hit. As a result, the Army decided to use chemical riot agents to drive the enemy from their tunnels and into the open.

During Operation PERSHING during 1967, the 1st Cavalry Division used Chinooks to drop a total of 30,000 pounds of tear gas agents on enemy positions. The CH-47s used a simple locally fabricated fusing system which was mounted on a standard fifty-five gallon gas drum. The drums were rolled off the rear cargo ramp and the fuse was armed by a static line, exploding a pre-determined distance off the ground and spreading the tear gas over a wide area. During this same time period, napalm drums were fused and dropped in a similar manner. A single CH-47 could lay down two and one half tons of napalm in a single drop, making it a highly effective weapon against Viet Cong tunnels.

All U.S. Army Chinooks committed to Vietnam were finished in overall Olive Drab paint schemes. Soon after its initial deployment to Vietnam the early gloss finish was flattened and all high visibility markings were subdued, becoming Flat Black. Additionally, the Star and Bar national insignia was removed to deny the Viet Cong a convenient high visibility aiming point. Most CH-47B and CH-47Cs were delivered from the factory in the Flat Olive Drab scheme. Like most aircraft assigned to combat duty, Chinooks displayed an abundance of unit and personal markings. Fuselage spines, rotor pylon top surfaces, and upper rotor blade surfaces were usually painted White or Orange as a recognition aid to orbiting support aircraft. The color and dimensions of the highlighted areas were usually based on command directives and unit policies.

The Chinook logged an enviable record of performance in Vietnam as evidenced by the following statistics — 1,182,000 flight hours (996,000 in combat); 2.6 million sorties; 4.5 million tons of cargo; 8.5 million passengers transported; all while maintaining a seventy-one percent aircraft availability rate. A total of 170 Chinooks were lost in Vietnam to all causes (combat and accidents).

A CH-47C of C Company, 159th ASHB, 101st Airborne Division, moves a heavy sling load of supplies near Chu Lai, Vietnam, during 1969. The crewman in the open hatch watched for signs of enemy activity and kept an eye on the external load. (Dave Grieger)

A CH-47C of the 196th Aviation Company, 18th Airborne Corps at Fort Bragg, North Carolina lifts a pair of M102 Howitzers. Chinoook are the primary platform used to emplace artillery units into remote fire bases. (U.S. Army)

One part of the XM-24 armament system was a pintel mounted M60D machine gun in the port escape hatch. The canvas bag attached to the gun collected the ejected shell casings and prevented them from being blown out of the aircraft and damaging the fuselage. (Bob Steinbrunn)

A crewman of a CH-47A fires the ramp mounted M60D machine gun while wearing a "monkey harness" gunner's belt designed to keep him attached to the aircraft during violent maneuvers. (The Squadron/Signal editor has had the unpleasant experience of being tossed out of a CH-34 while on a fence patrol mission and dangling from a "monkey harness.") (U.S. Army)

CH-47 Armament

Starboard Crew Door Gun Mount — M60 D Machine Gun, Quick Disconnect Mount, Ammo Box

Port Side Escape Hatch — Ejected Casings Bag, Pintel Mount, Ammo Box

Cargo Ramp Mount — Ejected Casings Bag, Bipod, Ammo Box

This CH-47A is transporting a sling loaded Bell UH-1 Huey of the 175th AHC Outlaws over the Vietnamese coast. The Huey's rotors have been removed and are carried as internal cargo on the CH-47. (Boeing Vertol)

A CH-47A transports a damaged CV-2 Caribou transport to a repair base in Vietnam. The CV-2 has had the outer wing panels, engines, and tail surfaces removed to reduce the total weight being lifted by the Chinook. (Boeing Vertol)

The CH-47 was the primary recovery vehicle in Vietnam for downed aircraft and helicopters. This CH-47B of the 180th ASHC is transporting a damaged UH-1C Huey back to its base at Can Tho, Vietnam during 1971. (Tom Hansen)

Many of the aircraft Chinooks recovered in Vietnam, such as this UH-1H Huey of the 176 AHC, went down in hostile territory and the crews could not take the time to properly prepare the aircraft for sling loading. (Dave Grieger)

Chinooks in Vietnam were called upon to haul a countless variety of cargo, including some decidedly non-military items. This CH-47A of the 178th ASHC loads a Vietnamese farmer's ox cart for delivery to a resettlement village. (Boeing Vertol)

FM Antenna Systems

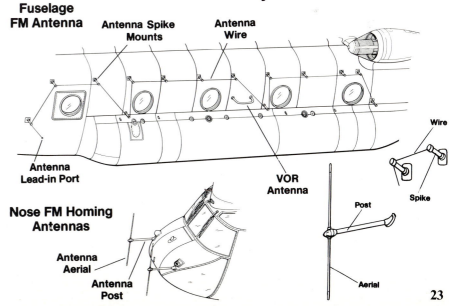

WAR WAGON, a CH-47A of B Company, 228th ASHB, 1st Cavalry Division, rests in its revetment at Bear Cat, Vietnam, during 1971. The insignia consisted of a White skull with a Gray helmet over a Black cross, surrounded by Yellow/Orange flames. The name *WAR WAGON* was in Black. (Author)

This CH-47B, named *Easy Rider*, had part of its chin window shot away while on a mission over Vietnam. The copilot's cyclic and thrust control stick can be seen just inside the cockpit through the removed copilot's door. (via Pete Harlem)

Chinooks in Vietnam were often painted with colorful personal, as well as, unit markings. *The Virgin Hunter* was a CH-47A of C Company, 159th ASHB, 101st Airborne Division. (Bob Chenoweth)

Troops of the 199th Light Infantry Brigade on the ground hide their faces from the dust and dirt thrown up by the rotor down wash of this CH-47A landing at a forward fire base in Vietnam during 1968. (Bob Krenkel)

This CH-47A of the 330th Transportation Company kept a running tally of the aircraft it recovered on the forward fuselage. There are 167 Black silhouettes around the crew door, representing the various types of aircraft recovered as of November of 1967. (Terry Love)

This CH-47A is being used to emplace a mobile radar unit on a tower at a fire support base just south of Siagon during 1968. Chinooks were widely used for moving such outsized cargo into tight areas. (Bob Krenkel)

A CH-47A prepares to lift a sling load of cargo at a fire base in support of the 199th Light Infantry Brigade in Vietnam during 1968. The dark circle at the rear of the fuselage is an open cabin window which was used as a gun port by the troops on board. (Bob Krenkel)

Chinooks quickly became the primary means of support for remote artillery fire bases in Vietnam. This CH-47A of B Company, 228th ASHB, 1st Cavalry, picks up a sling load of spent shell casings and water bladders from a fire base during 1968. (U.S. Army)

These racks being loaded aboard a CH-47 of the 1st Aviation Brigade at Di An, Vietnam, in June of 1968 were custom built to handle CS (tear gas) barrels. The barrels were rolled out of the back of the Chinook and exploded at a pre-set altitude to disperse the tear gas over a large area. (U.S. Army)

The transportation of villagers and their belongings was normal duty for CH-47 Chinooks during the Vietnam war. On such flights, the normal passenger load of the CH-47 was often far exceeded. (Boeing Vertol)

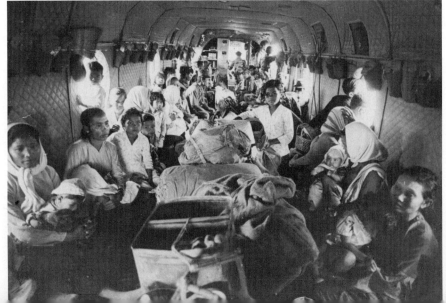

A CH-47A of the 178th ASHC carries a pontoon bridge section suspended from its external cargo hook. Chinook and Skycrane helicopters were the only Army helicopters in Vietnam capable of transporting and emplacing these pontoon bridges. (Boeing Vertol)

A ship's crewman directs this CH-47A onto the helicopter landing pad aboard a U.S. Navy supply ship off the coast of Vietnam. Chinooks were often used to move large quantities of supplies from ships to bases on shore. (Boeing Vertol)

The pilot of this CH-47A must maintain a steady hover while an Army recon team is recovered from the Vietnamese jungle. The stability of the Chinook in the hover was greatly appreciated, and few other helicopters could have carried out this mission in Vietnam's hot and humid climate. (Boeing Vertol)

This faded and war weary CH-47A at Bear Cat, Vietnam, during 1971 carries the insignia of B Company, 228th ASHB, 1st Cavalry, on the rear rotor plyon. One of the identifying features of the CH-47A is the tapered trailing edge of the rear rotor pylon. (Author)

180th ASHC crew members survey the damage to their CH-47A after it was hit while conducting a sling load delivery flight during 1968. The Black smudge of a shell hole can be seen just behind the pilot's door. (Bob Steinbrunn)

A CH-47C of the 179th ASHC, Shrimp Boats, refuels on the ramp at Vinh Long, Vietnam, during 1971. The Chinook carries a White *SUPER C* insignia near the top of the rear rotor pylon. The Red, White, and Blue flash on the pylon trailing edge identifies the 52nd CAB as this Chinook's parent command. (Tom Hansen)

A/ACH-47A

During 1964 the Army began looking for a new helicopter gunship to replace the Bell UH-1B Huey gunships which had been operating in Vietnam since 1962. The Army felt that there was a need for a heavily armed and armored helicopter gunship; and, although this concept was highly favored by the Army, it was opposed by the Commander in Chief, Pacific (CINCPAC) and Joint Chiefs of Staff (JCS), who both recommended expanding the use of fixed-wing aircraft in the ground support role.

During 1965, the Army held a competition to select an in-production helicopter for conversion to the dedicated gunship configuration. Five entries were evaluated before the Bell AH-1G Cobra was finally selected to be the Army's primary helicopter gunship. The Cobra was intended to be an interim aerial heavy weapons platform until the Lockheed AH-56 Cheyenne became operational. In the event, the AH-56 program met with economic problems and was cancelled.

The Army also expressed an interest in a large, heavy gunship, which could carry more firepower than the AH-1G, and ordered a prototype of Boeing Vertol's proposed A/ACH-47A (Armed and Armored CH-47A). On 23 June 1965, Boeing Vertol submitted a formal proposal for modifying eleven production CH-47A helicopters, ten for operational use and a single aircraft for testing. The Army accepted the proposal and a contract was signed on 30 June 1965. On 2 July the test aircraft (64-13145) was diverted from the production line and began conversion as the prototype A/ACH-47A.

The load carrying capability of the CH-47A allowed Boeing engineers a great deal of flexibility in selecting the armament systems, ammunition loads, and armor protection to be installed on the A/ACH-47A. Numerous systems were considered, including manned gun turrets, Minigun installations, cannons, grenade launchers, rockets, and a short stub wing which could mount various combinations of weapons. After careful study the optimum blend of weapons was finally selected and installed on the prototype.

The prototype A/ACH-47A made its first flight on 6 November 1965, with the official rollout ceremony being held four days later on 10 November. After company tests, the prototype was delivered to the Army during December. After extensive testing of the prototype, the Army ordered three production aircraft, which were delivered on 10 February 1966 (the initial contract for eleven aircraft had been reduced to four aircraft because of budget limitations). The fourth A/ACH-47A was retained by Boeing Vertol for additional company testing.

During May of 1966, all four aircraft were deployed to Vietnam, where they formed the 53rd Aviation Detachment of the 1st Cavalry Division. Better known as the *Guns A-Go-Go*, the unit became operational on 26 June 1966 and began an intensive combat evaluation of the Chinook gunship. During their combat evaluation, the Chinooks were used in support of the 1st Cavalry Division, the Royal Australian Task Force, IV Corps, and the 1st and 25th Infantry Divisions.

Equipped with uprated 2,850 shp Lycoming T55-L-7 turbine engines to boost performance, the flight performance of a fully loaded Chinook gunship was equal to that of a standard CH-47A. In combat the aircraft normally carried a crew of eight plus full fuel and ordnance loads, giving it a maximum gross weight of 31,600 pounds and a takeoff gross weight of 29,770 pounds. Fully loaded the aircraft could hover at 4,600 feet; the cruising speed was 125 knots and 130 knots with its ammunition load expended. With a full fuel load of 619 gallons, the A/ACH-47A had a mission duration of some two hours.

As part of the conversion to the gunship role, the aircraft had the troop seats, cargo hook, winch, heater, and soundproofing removed to cut down on unnecessary weight. Armored seats for the pilot and copilot (capable of withstanding up to .50 caliber

Two of the four A/ACH-47As (64-13149 and 64-13145) gunships at Aberdeen Proving Grounds during early 1966. Both aircraft carry high visibility markings and a Gloss Olive Drab finish. The aircraft in the foreground (13145) is equipped with engine intake screens which later became standard on all CH-47s. (U.S. Army)

Nose Development

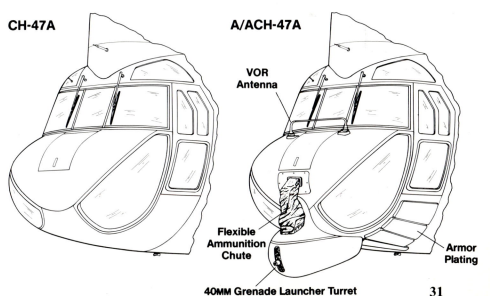

The flexible gun sight for the M-5 system was located at the copilot's station. The pilot and copilot sat in armored seats to give them better protection from enemy small arms fire. (U.S. Army)

rounds), weapons hardpoints, and external armor protection was installed with a total of 2,681 pounds of armor being added to the airframe. The armor protection included bolt-on plates installed under the lower nose section and additional plates attached to the rotor pylons.

The A/ACH-47A was armed with an XM-5 40MM grenade launcher in a turret under the nose which was controlled by the copilot. The XM-5 system carried 500 rounds of ammunition and could fire some 250 rounds per minute (rpm) at a range of 1,500 meters. The A/ACH-47A was fitted with stub wings mounted to the fuselage sides over the forward landing gear. These stub wings carried an M-24 20MM cannon mounted on each end plates. Some 800 rounds of ammunition for the cannons was stored in the fuselage and was fed to the guns through flexible chutes. The M-24 had a rate of fire of 800 rpm and a range of 5,200 meters. Pylons and bomb shackles, which could accept either an XM-159 (19 shot) 2.75 inch rocket pods or a 3,000 rpm XM-18 Minigun pod, were mounted under each stub wing. These weapons were aimed by maneuvering the entire aircraft and were controlled by the pilot using a fixed gun sight.

Special mounts were also fitted for crew-served M-2 .50 caliber machine guns at five positions — forward starboard side, forward port side escape hatch, two rear fuselage waist positions, and the rear cargo ramp. The rear overhead cargo door was removed to give a clear field of fire for the rear gun and the fuselage window opening at each gun position was enlarged to allow for a greater field of fire. The pintle gun mounts at each gun position could also accept the 7.62MM machine gun in place of the .50 caliber machine gun, however, the .50 caliber gun was usually retained because of its superior weight of fire. A total of 4,000 rounds of .50 caliber ammunition was carried, stored in containers alongside each gun position and fed to the guns through flexible chutes. The overlapping fields of fire gave the A/ACH-47A 360° coverage of a landing zone. The total weight of the system; guns, mounts, and ammunition was 3,000 pounds.

A/ACH-47A Armored Pilot's/Co-Pilot's Seats

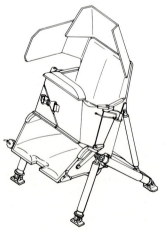

Closed

Open

An A/ACH-47A undergoes maintenance on the nose-mounted M-5 armament system at An Khe, Vietnam, during September of 1966. The curved plates under the nose and the plates on the sides of the forward rotor pylon are externally mounted armor plate. The unit insignia is carried on the front of the forward rotor pylon. (U.S. Army)

The starboard forward .50 caliber machine gun position on a A/ACH-47A. The ammunition was fed from the ammunition bin at the right through flexible chutes to the gun. The curved collector bin under the gun caught ejected shell casings and fed them into the bag under the gun. (U.S. Army)

A/ACH-47A Armament

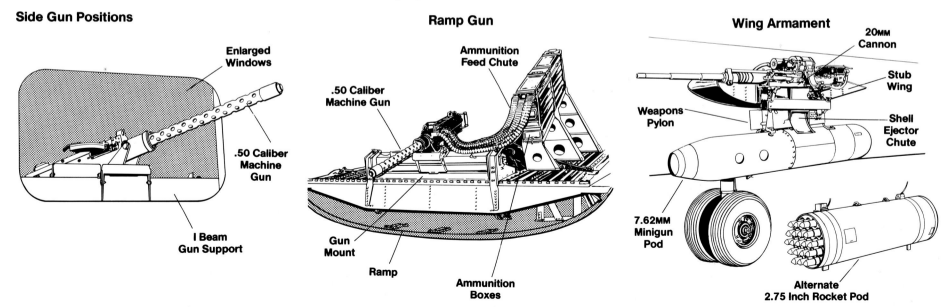

Side Gun Positions — Enlarged Windows, .50 Caliber Machine Gun, I Beam Gun Support

Ramp Gun — .50 Caliber Machine Gun, Ammunition Feed Chute, Gun Mount, Ramp, Ammunition Boxes

Wing Armament — 20mm Cannon, Stub Wing, Shell Ejector Chute, Weapons Pylon, 7.62mm Minigun Pod, Alternate 2.75 Inch Rocket Pod

While largely successful in combat, the unit was plagued with a series of accidents and battle damage, resulting in the loss of three Chinooks by January of 1967. At that time, the unit disbanded and the sole remaining A/ACH-47A was turned over to the 1st Aviation Brigade. The mix of weapons, the heavy volume of firepower, and endurance had allowed the gunship Chinooks to neutralize every enemy position detected and attacked during their evaluation. Although anything but graceful, the A/ACH-47A had a tremendous effect on the morale of the troops they were supporting. The program was cancelled, however, when the 1st Cavalry Division submitted their evaluation report which stated that the A/ACH-47A required an inordinate amount of maintenance support and that further conversion of CH-47s to the gunship configuration would reduce the number of available (and badly needed) lift aircraft.

M-24 20MM cannons were mounted on the ends of both stub wings and fed through flexible chutes from an ammunition bin inside the cabin. The bomb rack below the gun was capable of carrying either a nineteen shot 2.75 inch rocket pod or a minigun pod. (U.S. Army)

One of the evaluation A/ACH-47A gunships rests among 1st Cavalry Chinooks at the 228th Aviation Battalion parking area at An Khe, Vietnam during September of 1966. During the evaluation period, the aircraft were attached to the 228th for support and maintenance. (U.S. Army)

This A/ACH-47A Chinook gunship is armed with pylon mounted 20MM cannons, an XM-159 2.75 inch rocket pod on the port pylon, and an XM-18 7.62MM minigun pod on the starboard pylon. This Chinook gunship still carries the early Gloss Olive Drab finish with high visibility markings. (U.S. Army)

External armor plates were added under the port side of the forward fuselage to protect the pilots and flight control linkage from enemy ground fire. At the left is the M-5 40MM ammunition chute loaded with 40MM rounds. (U.S. Army)

The ramp .50 caliber machine gun position gave the gunner a wide field of fire behind and below the aircraft. The ejected shell casings were collected in the bag at the gunner's feet. The gunner is attached to the cabin wall with a gunner's belt to prevent him from being thrown from the aircraft. (U.S. Army)

International Military Chinook

The first Chinooks to be exported were thirty-nine CH-47As and CH-47Bs that were delivered to the South Vietnamese Air Force (VNAF) between late 1969 and 1972. These aircraft served in the VNAF throughout the war and reportedly some of the surviving aircraft were also used by the North Vietnamese after the fall of South Vietnam. Thailand received four CH-47As which were delivered direct from U.S. Army stocks.

During 1968, Elicotteri Meridionali SpA (EMSA), part of the Italian Agusta Aerospace Group, obtained a license from Boeing Vertol to produce the CH-47C in Italy. Under the terms of the license agreement Agusta was permitted to pursue export sales of the CH-47C in the Middle East and Mediterranean markets. The initial order for the Italian built CH-47C was twenty-six aircraft for the Italian Army. This contract was followed by an order for ninety-five aircraft for Imperial Iranian Army Aviation (93) and the Imperial Iranian Air Force (2).

To help the Italians avoid protracted delays in the delivery of these aircraft while the production lines were being set up, Boeing Vertol supplied the first two Italian Army Chinooks and thirty-eight of the Iranian Chinooks (under the designation Model BV-219). EMSA's production is currently limited to the BV-414, which is based on the U.S. Army's CH-47C. Besides producing Chinooks for Iran and the Italian government EMSA has also supplied Chinooks to Morocco (6), Egypt (25), Libya (20), and Greece (10). When the Iranian Revolution of 1981 caused the U.S. government to embargo further deliveries of military equipment to Iran, the remaining fourteen aircraft on the Iranian contract were diverted and became part of the Egyptian contract. To date, EMSA has produced more than 200 Chinooks.

The first export sale negotiated by Boeing Vertol was an order for twelve Model BV-165s (CH-47Cs) placed by the Australian government during 1972 for the Royal Australian Air Force (RAAF). These aircraft, designated A-15 Chinooks by the RAAF, were fitted with a crash-worthy fuel system. Additionally the power ratings of the engines was increased to allow safe operations at gross weight under the hot and high altitude conditions found in Papua New Guinea. RAAF Chinook deliveries were completed during 1973 and the aircraft were used to reactivate No 12 Squadron at Amberley, Queensland in March of 1974.

Spain has been a Boeing customer since 1972 when an order for ten Chinooks, designated the BV-176 (CH-47C), was placed on behalf of the Spanish Army. The first of these was delivered on 17 December 1972 and by September of 1973, all ten Chinooks had been delivered and were operational with Helicopter Unit 5 at Los Remedios. The Model 176 while externally identical to the CH-47C, featured a more advanced automatic flight control system, a quick-change cabin ferry fuel system, and a removable VIP cabin module. On 30 October 1980, the Spanish Army ordered three CH-47D Chinooks which were also assigned to Helicopter Unit 5. A follow-on order in December of 1984 for six additional CH-47Ds has increased the Spanish Chinook fleet to eighteen aircraft.

During 1973, Boeing Vertol received an order from the Canadian Armed Forces for eight Chinooks, designated the BV-173. These aircraft were basically similar to the CH-47C and received the Canadian designation CH-147. The CH-147 introduced a number of improvements over the CH-47C, including an increase in seating capacity to forty-four (by the addition of eleven center aisle seats), a power drive system for the rear loading ramp, and an inflatable water dam for sustained on water operations. In addition, the avionics and autopilot were improved and the optional rescue hoist became a standard

This CH-47A, assigned to the Vietnamese Air Force (VNAF) during late 1971 carried no armament and had the lower copilot's window covered over with armor plate. (Boeing Vertol)

One of four CH-47A Chinooks that were supplied to the Royal Thai Air Force during the Vietnam war directly from U.S. Army stocks. It is believed that three of these aircraft are still in service. (Boeing Vertol)

feature, installed over the front starboard crew door. The CH-147 was the first Chinook to be equipped with a 28,000 pound capacity external cargo hook. With an uprated transmission and improved 3,750 shp Lycoming T55-L-11C engines, the aircraft could be operated at gross weights up to 36,000 pounds, compared with 28,500 pounds for the standard CH-47C. The first CH-147 crashed on its delivery flight in October of 1974; however, the eight surviving Chinooks were safely delivered and are assigned to Nos 447 and 450 Transport Squadrons. Canadian Chinooks are often equipped with ski landing gear for operations in snow and ice. The skis are designed to attach directly to the wheel mounts and require no special modifications to the aircraft or the landing gear.

After two successive Chinook orders were cancelled due to economic problems, Great Britain was finally able to fund a Chinook purchase for the Royal Air Force in January of 1978, placing an order for thirty-three Chinooks under the designation Chinook HC Mk 1 (BV-352). The first Chinook HC Mk 1 was delivered to the RAF on 30 September 1980 and became operational with the No 240 Operational Conversion Unit on 22 November. Three additional aircraft were ordered during 1982 to replace those lost aboard the ATLANTIC CONVEYOR which was sunk during the fighting in the Falklands. During late 1981, the RAF's first all-Chinook squadron became operational and by March of 1986 forty-one HC Mk 1s had been delivered.

The HC Mk 1 uses the same improvements specified for the Canadian Chinook along with a triple external cargo hook system, enhanced British-made navigation and communications avionics equipment, a fuel fire suppression system, and zoned cockpit lighting. Power is provided by a pair of 3,750 shp T55-L-11E engines driving sixty foot diameter, thirty-two inch chord fiberglass rotors. A rotor brake was installed to assist in rough weather starting and allowed the crew to quickly fold the blades to conceal the aircraft when operating in the field. Other British specified equipment included a Night Sun searchlight system, downward facing floodlights, and identification lights.

Although the Argentine Government originally ordered thirty-five Chinooks, only five (designated the BV-308 and BV-309, for the Argentine Air Force and Army respectively) were delivered during 1980. Based on the CH-47C, these Chinooks serve as Antarctic logistic/rescue aircraft and are fitted with weather radar in a modified nose section and duplex inertial navigation systems. These aircraft also have provision for internal fuel tanks carried in the cargo compartment, when installed, these tanks increase their range to 1,265 miles.

Boeing Vertol's newest international customer is Japan which, since 1985, has been receiving major sub-assemblies from Boeing Vertol as part of a co-production and assembly program to be undertaken in Japan. The Japanese firm of Kawasaki Heavy Industries is currently scheduled to produce fifty-five Chinooks, under the designation CH-47J with forty aircraft scheduled for delivery to the Japanese Ground Self-Defense Force (JGSDF) and the remaining fifteen scheduled to be delivered to the Air Self-Defense Force (JASDF). Under a three-phase program, begun during 1986, Boeing Vertol ships Chinook major assemblies to Kawasaki, which completes assembly of the aircraft and adds customer specified equipment. On 4 July 1986, Kawasaki successfully flight tested the first Japanese assembled aircraft and deliveries to the JGSDF began during late 1986. With the exception of customer specified equipment the CH-47J is identical to the CH-47D.

Other export customers for the CH-47 have included Israel, South Korea, Syria, and Tanzania.

This Italian Army CH-47C landing at an Italian base was built in Italy under license. The aircraft is camouflaged with Dark Gray and Dark Green uppersurfaces over Medium Gray undersurfaces. (Sergio Lion via Terry Love)

Ninety-five Chinooks were slated for delivery to Iran before the United States embargoed further deliveries of military equipment to Iran. In the event, only sixty-seven aircraft were actually delivered. The camouflage is Sand and Tan over Light Gray. (via Terry Love)

This CH-47C (A15) of the Royal Australian Air Force is camouflaged with Tan and Medium Green uppersurfaces over Flat Black undersurfaces. The insignia is a low visibility type and consists of a Black Kangaroo silhouette on the center fuselage without the normal Blue and White roundel. (Mike Macgowan via Terry Love)

One of twenty Italian-built Model 414 Chinooks that were delivered to the Libyan Arab Air Force. Camouflaged in Sand and Dark Brown uppersurfaces over Light Gray undersurfaces, these Chinooks have since had the Red/White/Black roundel replaced by a solid Dark Green roundel. (Boeing Vertol)

This Italian-built CH-47 was purchased by Greece for the Helenic Air Force during 1982. The camouflage is Gloss Tan, Brown, and Dark Green uppersurfaces over Light Gray undersurfaces. (via Terry Love)

This Canadian Armed Forces CH-147 is an improved version of the U.S. Army's CH-47C. The rescue hoist over the cabin crew door is a standard feature on all Canadian Chinooks. (Terry Love)

Optional Equipment

Rescue Hoist (Standard on Canadian CH-47)

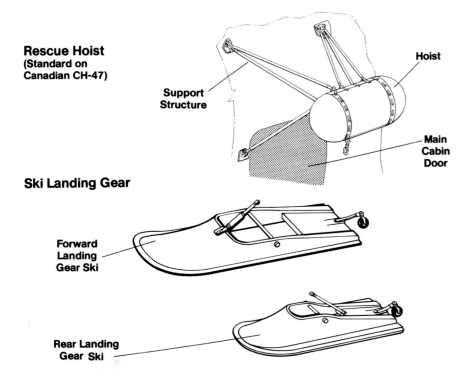

Ski Landing Gear

A ski-equipped Canadian Armed Forces CH-147 Chinook (CH-47C) sling loads a pair of 105MM artillery pieces during a Canadian exercise. The skis can be easily detached for operations from hard runways. (Boeing Vertol)

A Chinook HC Mk 1 (ZA680) of No 18 Squadron, Royal Air Force lifts a nine ton armored Combat Recon Vehicle (CRV) at the primary Chinook base at Odiham, England. (Boeing Vertol)

A Spanish Army Chinook flies low over the Spanish countryside. The Spanish Army operates eighteen Chinooks as part of Helicopter Unit Five and has ordered a number of CH-47Ds to augment this force. (Boeing Vertol)

The Royal Air Force received their first Chinooks during the late 1980s. The tunnel covers and rear rotor pylon doors are open on this Chinook HC Mk 1 for maintenance of the drive shaft. This Chinook has also had one of the cabin windows replaced with a bulged observation window. (via Terry Love)

The CH-47J is based on the U.S. Army CH-47D and is coproduced by Boeing Vertol and Kawasaki Heavy Industries, being assembled in Japan for both the Japanese Air and Ground Self Defense Forces. (Boeing Vertol)

This Model BV-308 of the Argentine Air Force was based on the CH-47C. Argentine Air Force Chinooks are used primarily to support the Antartic mission and are finished in a high visibility scheme of International Orange over Light Gray. This BV-308 has been modified to carry a weather radar in an extended nose cone. (Boeing Vertol)

Nose Development

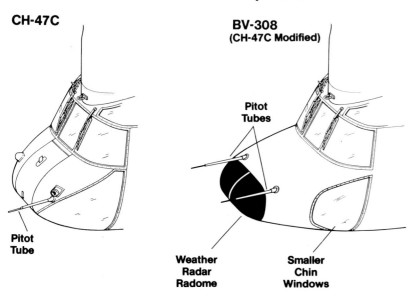

CH-47C

Pitot Tube

BV-308 (CH-47C Modified)

Pitot Tubes

Weather Radar Radome

Smaller Chin Windows

CH-47D

Faced with an aging inventory of CH-47 Chinooks during the mid 1970s, the Army determined that the most cost effective method to maintain the required numbers of Chinooks needed to meet its medium lift requirements would be to modernize the existing CH-47 fleet. The resulting upgrade program calls for the modernization of 436 Chinooks by 1993, along with procurement of eighty-two new production aircraft under the designation CH-47D. Although nearly identical externally to the earlier CH-47C, the CH-47D is completely rebuilt. Each early CH-47 in the program is stripped to the basic airframe and totally refurbished with state-of-the-art equipment and electronics. The modernization is so complete that CH-47As that have been through the program are not only redesignated CH-47Ds, they are also assigned new serial numbers.

The CH-47D prototype was a rebuilt CH-47A (65-8008) which made its first flight on 11 May 1979. Production began during 1980 with the first aircraft to complete the program being delivered to the 159th Assault Helicopter Battalion (AHB), 101st Airborne Division at Fort Campbell, KY on 28 February 1983. Other units that have received CH-47Ds include the 24th Infantry Division, 9th Infantry Division, and 82nd Airborne Division.

Power for the CH-47D is derived from a pair of 3,750 shp Lycoming T55-L-712 turbine engines, capable of producing 4,075 shp for takeoff. The rotor blades fitted to the CH-47D are of fiberglass and, although retaining the same diameter as those on the CH-47C, have a thirty-two inch chord (twenty-four inch chord on the CH-47C). The greater rotor chord gives the CH-47D much greater lift and maneuverability, especially under hot and high conditions. The fiberglass rotors are also better able to withstand damage and are capable of absorbing hits from up to 23MM rounds without failure.

Other improvements were made to the CH-47D to enhance its capabilities and lower its operating costs. These include a new automatic flight control system, improved hydraulics, improved drive and electrical systems, updated avionics, a single-point high pressure refueling system, and cockpit lighting which is compatible with night vision goggles.

Survivability is also increased with the addition of lightweight armored crew seats, an AN/APR-39V radar warning set, an AN/ALQ-156 missile detection system, a crash resistant fuel system, and armor plate protecting the strengthened transmission and oil coolers. Antennas for the radar warning system are mounted on the nose and on the trailing edge of the rear rotor pylon, giving the system a full 360 degree coverage. To further increase survivability and give the CH-47D protection from shoulder fired infrared surface-to-air missiles (SAMs), an M-130 flare dispenser can be attached to the port side of the rear fuselage, where a small rectangular flare viewing window has been added. The CH-47D also incorporates a greater degree of systems redundancy than earlier Chinook models (the CH-47A had no systems redundancy) with all primary systems electrical and hydraulic systems having a back-up system.

The CH-47D has a maximum gross weight of 50,000 pounds. It can accommodate up to forty-four combat ready troops while carrying other loads externally for a maximum payload of up to 28,000 pounds. A multi-hook external cargo system permits hauling 'sling' loads at speeds in excess of 115 knots, compared to the previous limit of 40 to 60 knots for the CH-47A. The CH-47D employs a triple-hook arrangement, adding two additional freight hooks on the fuselage underside, each with a 20,000 pound capacity. The two additional cargo hooks are located one ahead of and one behind the main 28,000 pound capacity hook. One advantage of the triple hook system is that the CH-47D can transport external loads for three different destinations on a single mission.

With a maximum takeoff weight of 53,000 pounds, the CH-47D can climb at 3,130 feet per minute and hover at 17,000 feet. It has a maximum speed of 185 mph and a cruising speed of 159 mph. Its fuel capacity is slightly less than that of the CH-47C at 1,030 gallons.

This new production CH-47D at Fort Lauderdale, Florida reveals the wide-chord fiberglass rotors introduced on the CH-47D. The third cabin window has been replaced with a bubble, bulged observation window. (Dean Noel)

The YCH-47D prototype (30129) made its official roll-out in November of 1980. The aircraft was a completely rebuilt and modernized CH-47A, and served as one of the pattern aircraft for the CH-47D rebuild program. (R.R. Leader via Pete Harlem)

With the introduction of the CH-47D the Army now has a medium lift helicopter that is capable of battlefield support missions well into the late 1990s. The CH-47D is the first medium lift helicopter that can lift an M198 155MM howitzer, its eleven man crew, and thirty-two rounds of ammunition on a single mission. Previously to accomplish this same mission would have required at least two CH-47Cs.

To further increase the capabilities of the CH-47D the Army undertook a series of aerial refueling tests with a modified CH-47D. The aircraft had the fuel system modified to accept a thirty foot telescoping refueling probe which was mounted on the starboard side of the fuselage. During 1985 the system was flight tested and the Chinook completed thirty-five successful hook-ups behind an HC-130E tanker. Hook-ups were performed at altitudes up to 5,000 feet and at speeds up to 120 knots. During July of 1988 the first production CH-47D with the modified fuel system was delivered to the Army. Reportedly all further CH-47D conversions will also feature this fuel system and earlier aircraft will reportedly be retrofitted at a later date. This modification will enable the Chinook to self-deploy to Europe without having to use up cargo space in the main cabin with bulky ferry fuel tanks.

MH-47E

A specialized variant of the CH-47D has been proposed by the Army's Task Force 160 for use by Special Operations Forces such as Delta Force under the designation MH-47E. Reportedly, seventeen aircraft are currently on order with options for another thirty-four.

Although little has been released about the program, the MH-47E is described as a CH-47D fitted with uprated Lycoming 4,000 shp T55-L-714 engines, enlarged fuel tanks, a fixed in-flight refueling probe, Forward Looking Infrared (FLIR), radar, satellite communications equipment, a rescue winch mounted over the starboard crew door, and armament such as mini-guns or .50 caliber machine guns. Currently, the prototype MH-47E is expected to make its first flight in December of 1989, to be followed by a testing period of some seven to eight months.

Missions for the MH-47E would include covert Special Forces operations, combat rescue, deep penetration missions, and anti-terrorist operations under all weather conditions, day or night.

(Above) A C-5 transport can easily accommodate a CH-47D in its cargo bay without having to remove the engines, although the rotor pylons must be removed and carried as separate cargo. (Boeing Vertol)

(Right) A CH-47D uses two of the three external cargo hooks to lift a resupply container during April of 1980 Army exercise while a Bell AH-1G Cobra gunship hovers protectively nearby. (U.S. Army)

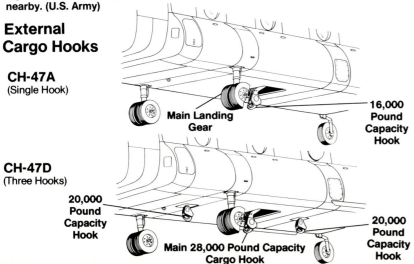

External Cargo Hooks

CH-47A (Single Hook)

CH-47D (Three Hooks)

Specifications
Boeing Vertol CH-47D

Rotor Span	60 feet
Length	51 feet
Height	18 feet 7.8 inches
Empty Weight	22,452 pounds
Maximum Weight	50,000 pounds
Powerplant	Two 3,750 shp Lycoming T55-L-712 turbine engines
Armament	Three M60D machine guns
Performance	
Maximum Speed	185 mph
Service ceiling	8,500 feet
Range	230 miles
Crew	Four

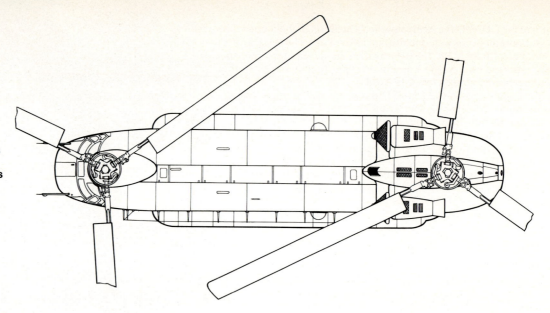
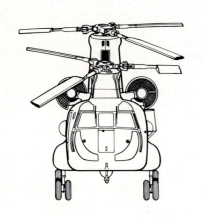
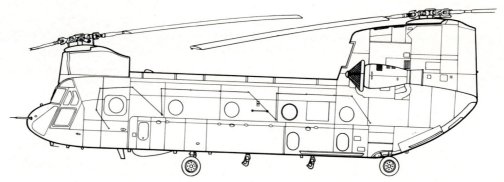
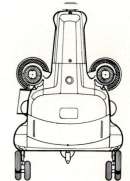
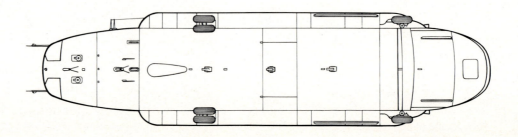

A CH-47D prepares to lift a medical module dolly at Hanau, Germany in September of 1983. The CH-47D had the pitot tube relocated on the port side of the nose and radar warning antennas mounted just forward of the upper portion of the large chin windows. (USAF)

This CH-47D was fitted with a telescoping refueling probe for aerial refueling tests during 1985. Although identified as a CH-47D, this aircraft has the extended nose and radome intended for the MH-47E. (Boeing Vertol)

Test Programs

The National Aeronautics and Space Administration operated at least one CH-47B for a series of joint Army/NASA tests during 1985. The aircraft was delivered to NASA from Army stocks and was fitted with a variety of experimental flight control systems which were under consideration for use on future helicopter programs.

BV-347

As part of the Heavy Lift Helicopter (HLH) program, Boeing Vertol developed the experimental BV-347 to test various concepts and components for use on future heavy lift helicopters. The prototype Model 347 was a converted CH-47A (65-7992) and made its first flight during 1972. The aircraft was extensively modified and featured a nine foot two inch stretch in the main cabin, retractable landing gear, four bladed rotors which were longer than standard CH-47 blades by two and one half feet, and a raised rear rotor pylon. The most unusual modification was the provision for a wing which could be mounted on the top center fuselage to augment the lift from the rotors. The prototype was also fitted with an advanced fly-by-wire control system and was powered by uprated Lycoming T55-L-11 engines giving it a maximum speed exceeding 180 knots. After the prototype completed the test program during 1975, it was retired and now rests at the Army Aviation Museum at Fort Rucker, Alabama.

This CH-47B carries the Blue and White scheme used by NASA aircraft. The aircraft was used by NASA during a series of joint NASA/Army tests of improved flight control systems during 1985. (Terry Love)

XCH-62A

Another project which also came about from the Army's HLH program was the Boeing Vertol Model 301, designated XCH-62A by the Army. The XCH-62A was designed as a possible replacement for the Sikorsky CH-54 Tarhe (Skycrane) helicopter and, although the XCH-62A HLH bears little resemblance to the Chinook, it evolved directly from the basic CH-47 design.

This ACH-62A was powered by three 8,079 shp Allison XT701-AD-700 turboshaft engines driving the tandem rotors. It was proposed that the XCH-62A have a crew of four and a detachable cargo compartment with optional seating for twelve troops. During construction of the prototype, a number of advanced technology components were tested. The airframe structure was of a bonded honeycomb construction, the control system was an advanced electronic fly-by-wire control system, and the rotor drive system was significantly improved.

One prototype was ordered (72-2012) during 1971 and was ninety-five percent complete when the program was cancelled during 1975. Boeing Vertol, however, continues to work on the aircraft, which remains in storage at the Boeing Vertol facility in Philadelphia. A proposed civil variant of the model BV-301, for possible airline use, was proposed by Boeing Vertol during 1978 under the company designation Model BV-307. This aircraft was to be almost double the length of a CH-47D and have a lengthened nose section to house a weather radar dish. To date no production contracts for the BV-307 have been forthcoming.

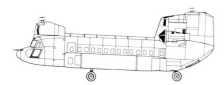

BV-307

BV-414
(CH-47D)

BV-307

(Above) The Boeing Vertol BV-347 Chinook was a one-of-a-kind test platform which incorporated a wing mounted above the lengthened fuselage. While normally fitted with four-bladed rotors, the aircraft was temporally fitted with three-bladed rotors while being demonstrated at Dulles Airport in May of 1972. (via Bob Pickett)

(Below) Boeing Vertol's XCH-62A Heavy Lift Helicopter (HLH) was designed as an advanced research aircraft and possible replacement for the Army Skycrane helicopter during the early 1970s. The ungainly looking aircraft was to be fitted with three 8,079 shp turbine engines and a detachable cabin capable of carrying twelve troops. (Boeing Vertol)

Commercial Chinooks

Boeing Vertol has developed a number of civil variants of the CH-47 which have gained wide acceptance within the civil/commercial market for both passenger and freight uses.

By 1986, civil variants of the Chinook were in service with British Airways Helicopters in Great Britain, Helkopter Service A/S of Norway, and ERA Helicopters and Columbia Helicopters (Alaska Helicopters) in the United States. The first civil Chinook, designated the Model BV-234, was first put into service by British Airways during 1979.

The Model 234 is available in four basic variants; the BV-234LR long range passenger/freight helicopter, the BV-234ER extended range passenger/freight helicopter, the BV-234UT utility helicopter optimized for tasks such as heavy construction work and resources exploration, and the BV-234MLR multi-purpose long range utility helicopter. The Model 234LR is a forty-four passenger aircraft with a gross weight of 45,200 pounds and features increased internal fuel capacity (2,100 gallons) carried in enlarged side sponsons, airliner type windows, and weather radar. The passenger seats are arranged in eleven four-abreast rows and, to increase passenger comfort, the cabin noise level was significantly reduced through the extensive use of interior soundproofing. The main cabin is easily and quickly convertible to a cargo configuration allowing the aircraft to be used for either freight or passenger missions. For safe all weather operations all BV-234 variants are equipped with dual backup avionics systems and a weather radar mounted in an extended nose. In anticipation of long range over-water operations, the civil Chinook under went a number of water tank trials which indicated that the Chinook will float with rotors either stopped or turning in as much as thirty foot high waves.

The BV-234ER extended range Chinook has provision for either one or two internal auxiliary fuel tanks installed in the cabin. In this configuration the number of seats is reduced to thirty-two (single fuel tank) or seventeen (dual tanks) depending on customer specifications. With two tanks, the BV-234ER has a range of 1,008 miles.

The BV-234UT has the external fuel tanks removed and the side fairings deleted. The cabin is configured with two drum shaped tanks mounted side by side in the front of the cabin. This variant is intended for utility freight operations and can carry up to 28,000 pounds on its single external cargo hook.

The last commercial variant, the BV-234MLR, is similar to the BV-234LR but has the passenger interior replaced by a cargo/utility interior.

Boeing Vertol is actively researching larger variants of the BV-234, one of which has a stretched fuselage that is some ten feet longer than the basic BV-234 and can carry sixty-eight passengers.

British Airways was an early commercial operator of the civil Model 234 Chinook. The commercial Chinook has extensively internal modifications to adapt it to the passenger role and airliner type windows along the fuselage sides. (Helicopter Magazine)

BV-234 Civil Chinook

CH-47D

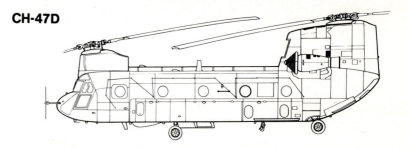

BV-234LR

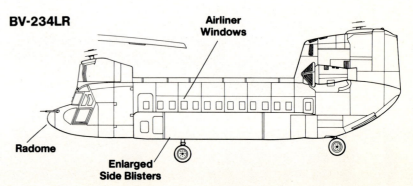

HeliTour Alaska Chinooks have a weather radar installed in the extended nose cap and are painted in a striking Red and White scheme. Besides flying tourists, this company also flies freight into remote Alaska locations. (HeliTour Alaska)

Along with enlarged side sponsons and improved electronics, a number of civil Chinooks feature a rescue hoist over the main cabin door. This Chinook of HeliTour Alaska has a cabin heater air intake installed just forward of the hoist. (HeliTour Alaska)

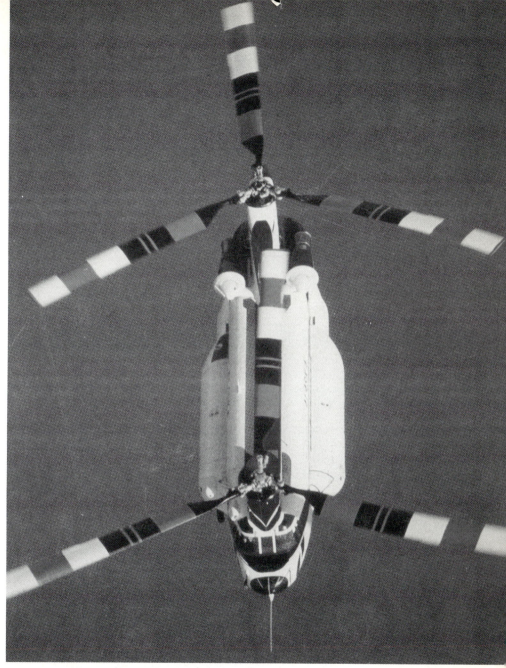

This Columbia Helicopters BV-234 reveals the enlarged side fuselage sponsons that house larger fuel tanks. This Chinook is overall White with Red trim and Black, Yellow, and White striped rotors. (Columbia Helicopters)

AIR SUPPORT IN VIETNAM
From squadron/signal

1014

1047

1065

1070

1075

1077

1087

1089

squadron/signal publications